T0146312

A Repair Network Concept for Air Force Maintenance

Conclusions from Analysis of C-130, F-16, and KC-135 Fleets

Robert S. Tripp, Ronald G. McGarvey, Ben D. Van Roo,
James M. Masters, Jerry M. Sollinger

Prepared for the United States Air Force
Approved for public release; distribution unlimited

PROJECT AIR FORCE

The research described in this report was sponsored by the United States Air Force under Contract FA7014-06-C-0001. Further information may be obtained from the Strategic Planning Division, Directorate of Plans, Hq USAF.

Library of Congress Cataloging-in-Publication Data

A repair network concept for Air Force maintenance : conclusions from analysis of
  C-130, F-16, and KC-135 fleets / Robert S. Tripp ... [et al.].
    p. cm.
  Includes bibliographical references.
  ISBN 978-0-8330-4804-2 (pbk. : alk. paper)
  1. Airplanes, Military—United States—Maintenance and repair—
Management—Evaluation. 2. United States. Air Force—Equipment—Maintenance
and repair—Management—Evaluation. 3. Hercules (Turboprop transports)—
Maintenance and repair—Management—Evaluation. 4. F-16 (Jet fighter plane)—
Maintenance and repair—Management—Evaluation. 5. KC-135 (Tanker
aircraft)—Maintenance and repair—Management—Evaluation. I. Tripp, Robert S.,
1944–

UG1243.R46 2010
358.4'1621—dc22

                                                                    2009049222

The RAND Corporation is a nonprofit research organization providing objective analysis and effective solutions that address the challenges facing the public and private sectors around the world. RAND's publications do not necessarily reflect the opinions of its research clients and sponsors.
RAND® is a registered trademark.

U.S. Air Force photo by Airman 1st Class Franklin J. Perkins

© Copyright 2010 RAND Corporation

Permission is given to duplicate this document for personal use only, as long as it is unaltered and complete. Copies may not be duplicated for commercial purposes. Unauthorized posting of RAND documents to a non-RAND Web site is prohibited. RAND documents are protected under copyright law. For information on reprint and linking permissions, please visit the RAND permissions page (http://www.rand.org/publications/permissions.html).

Published 2010 by the RAND Corporation
1776 Main Street, P.O. Box 2138, Santa Monica, CA 90407-2138
1200 South Hayes Street, Arlington, VA 22202-5050
4570 Fifth Avenue, Suite 600, Pittsburgh, PA 15213-2665
RAND URL: http://www.rand.org/
To order RAND documents or to obtain additional information, contact
Distribution Services: Telephone: (310) 451-7002;
Fax: (310) 451-6915; Email: order@rand.org

Preface

For more than 15 years, the U.S. Air Force has continually engaged in deployed operations in Southwest Asia and other locations. Recent Office of the Secretary of Defense (OSD) planning guidance directs the services to plan for high levels of engagement and deployed operations, although their nature, locations, durations, and intensity may be unknown. Recognizing that this new guidance might impose different demands on the logistics system, senior Air Force logistics leaders asked RAND Project AIR FORCE to undertake a logistics enterprise analysis. This analysis aims to identify and rethink the basic issues and premises on which the Air Force plans, organizes, and operates its logistics enterprise.

This monograph synthesizes the results of the initial phases of the logistics enterprise study. It describes an analysis of repair network options to support three series of aircraft: C-130, KC-135, and F-16. It assesses the effect of consolidating certain scheduled maintenance tasks and off-equipment component repair at centralized repair facilities. It also discusses an initial assessment of maintenance concepts that integrate wing-level and depot-level maintenance processes.

The Deputy Chief of Staff for Logistics, Installations and Mission Support, along with the Vice Commander, Air Force Materiel Command, sponsored this research, which was carried out in the Resource Management Program of RAND Project AIR FORCE under three projects: "Enterprise Transformation Management for AFMC Umbrella Project," "Global Materiel Management Strategy for the 21st Century

Air Force," and "Managing Workload Allocations in the USAF Global Repair Enterprise."

This monograph should interest logistics and operational personnel throughout the U.S. Department of Defense and those who determine logistics requirements.

This monograph summarizes work done as part of the RAND Project AIR FORCE Logistics Enterprise Analysis project. Other reports written as part of that project include *Analysis of Air Force Logistics Enterprise: Evaluation of Global Repair Network Options for Supporting the C-130* (Van Roo et al., forthcoming) and *Analysis of the Air Force Logistics Enterprise: Evaluation of Global Repair Network Options for Supporting the F-16 and KC-135* (McGarvey et al., 2009).

RAND Project AIR FORCE

RAND Project AIR FORCE (PAF), a division of the RAND Corporation, is the U.S. Air Force's federally funded research and development center for studies and analyses. PAF provides the Air Force with independent analyses of policy alternatives affecting the development, employment, combat readiness, and support of current and future aerospace forces. Research is conducted in four programs: Force Modernization and Employment; Manpower, Personnel, and Training; Resource Management; and Strategy and Doctrine.

Additional information about PAF is available on our Web site: http://www.rand.org/paf

Contents

Preface iii
Figures vii
Tables ix
Summary xi
Acknowledgments xv
Abbreviations xvii

CHAPTER ONE
Introduction 1
Motivation 1
Analytic Scope 2
Purpose 4
Approach 5
Organization of This Monograph 6

CHAPTER TWO
Requirement Determination and Allocation 7
Air Force Maintenance Practice 7
Determining Weapon-System Requirements and Workload 9
Allocating Workload Between Unit and Repair Network 10
Evaluating Efficiency 12
    Economies of Scale 12

CHAPTER THREE
Number, Location, and Size of Centralized Repair Facilities 21
Comparing Results of Multiple Mission-Design Series 25

Staffing Squadrons for Split Operations..29

CHAPTER FOUR
Effectiveness Analysis..31
Source of Effectivness...31
Integrating Processes..34

CHAPTER FIVE
Conclusions...37

Bibliography...39

Figures

1.1. Analytic Approach 6
2.1. Labor Economies of Scale for Centralized Repair Facility
      Isochronal Inspections 15
2.2. Labor Utilization Rates 16
2.3. C-130 Unit-Level Active-Duty and Reserve Personnel
      Requirements 18
2.4. C-130 Unit and Centralized Repair Facility Active-Duty
      and Reserve Personnel Requirements 19
3.1. Optimized C-130 Centralized Repair Facility Solutions 23
3.2. Current System Compared with C-130 Centralized
      Repair Facilities 25
3.3. Current System Compared with Centralized KC-135
      Repair Facilities 27
3.4. Current System Compared with F-16 Centralized Repair
      Facilities 28
4.1. Effect of Facility Size on Inspection Time 32
4.2. AFSOC and Little Rock AFB ISO-Inspection Flow Times 33
4.3. Number of C-130 Aircraft in ISO Inspection, as a
      Function of Network Type 33

Tables

2.1. C-130 Work Centers 11
3.1. Requirements for AD/AFRC Split Operations.................. 30

Summary

For more than 15 years, the U.S. Air Force has continually engaged in operations outside the United States. These operations have included humanitarian relief efforts, shows of force, support of allies, participation in coalition exercises, and a host of other missions. Current planning guidance from OSD indicates that this environment is likely to persist and directs the services to plan for high levels of such operations, although the specific nature, locations, durations, and intensity may be unknown. This is called the *steady-state security posture*. It depicts use of U.S. military capabilities different from that during the Cold War; the steady state is characterized by frequent deployments. Planning guidance still directs the services to develop capabilities to meet the requirements of major combat operations. At the same time, services are under pressure to operate more efficiently, to meet their mission responsibilities, and to contribute to joint expeditionary operations in Afghanistan and Iraq. Several logistics career fields have experienced serious stress, including security forces and civil engineering, in meeting these continuous deployment requirements. Therefore, the Air Force logistics leadership wishes to find more-efficient ways of supporting continuous aircraft deployments with fewer people. If this could be accomplished while providing the same or better level of effectiveness—e.g., aircraft availability—then some of the people freed up by more-efficient support could be reprogrammed into career fields that need it most, thereby making the Air Force more expeditionary.

To meet current and future aircraft deployment requirements, the Air Force has been using a logistics structure that was developed primarily to support the Cold War and to meet the requirements of large-scale combat operations. This structure provided for largely self-sufficient units that carried with them significant maintenance capabilities, stocks, and other resources, on the assumption that they would be cut off from transportation for long periods.

However, the Cold War logistics support structure may not be the best one to meet many of the demands of current and likely future requirements. The Cold War structure was tailored to support full-squadron deployments to a set of known locations and a specific operational tempo. However, recent engagements have called for different deployment concepts, such as those that employ only parts of squadrons and those that deploy forces to unexpected locations and for unknown durations. These partial-squadron deployments are referred to as *split operations* because they split a squadron into smaller deployment packages. These split-squadron operations require more maintenance personnel because the squadron operates at two locations, which requires more personnel to support. These additional personnel exceed authorizations, and the Air Force has decided not to fund the additional spaces. So, more-efficient ways are required to support split operations.

In addressing the inability of the Cold War structure to meet the Air Force's needs, the leadership saw an opportunity for a comprehensive strategic reassessment of the entire Air Force logistics system. Senior Air Force logistics leaders asked PAF to analyze the logistics enterprise to identify and rethink the basic issues and premises on which the Air Force plans, organizes, and operates its logistics enterprise from a total force perspective—including the active-duty Air Force, the Air Force Reserve, and Air National Guard.

At a fundamental level, the logistics enterprise strategy must answer the following three questions:

- What will the logistics workload be?
- How should the logistics workload be accomplished?
- How should these questions be revisited over time?

Research Approach

To answer these questions, we organized our research into four steps. First, we examined the OSD planning guidance to ascertain what the requirements for Air Force weapon systems are likely to be, and, from that, we calculated a logistics workload. Second, we determined what workload must be performed at the unit level—largely that associated with launching and recovering aircraft and removing and replacing parts or components. Third, we generated various network options for other workload with an eye to optimizing them from an efficiency and effectiveness standpoint. Our analysis considers every potential combination, from fully decentralized solutions to fully centralized ones. Finally, we tested the network designs for sensitivity to location and various policy considerations.

The complex nature of this project led us to approach it in phases. Thus far, we have analyzed the F-16, KC-135, and C-130 maintenance networks (see McGarvey et al., 2009; Van Roo et al., forthcoming). Subsequent analyses will examine other mission design series (MDS) (types and models of aircraft), such as strategic air lifters.

Findings

Our major overarching conclusion is that consolidating certain wing-level scheduled maintenance tasks and off-equipment component repairs is more effective and efficient than the current system, in which every wing has significant maintenance capabilities to support these activities. Consolidating inspections and back-shop maintenance is more efficient because it requires fewer people. (See pp. 12–19.) It is more effective because consolidation can speed the flow of aircraft through isochronal and phase inspections,[1] including associated com-

---

[1]   C-130 and other cargo aircraft undergo an isochronal inspection, based on calendar days since last inspection. F-16 (and other fighter aircraft) receive a phase inspection based on accumulated flying hours since last inspection. KC-135 undergo what is called a *periodic inspection*, defined as the earlier of a given number of calendar days or flying hours accumulated since last inspection.

ponent repairs, which means that fewer aircraft are tied up in maintenance processes at any given time, thus making more aircraft available to the operational community. (See pp. 31–34.) Consolidating backshop operations can provide immediate benefits and provide a good basis for integrating what are currently stovepiped intermediate- and depot-level processes, thereby opening up possibilities for even greater efficiencies and effectiveness.

Acknowledgments

Many people inside and outside the Air Force provided valuable assistance and support to our work. We thank Lt Gen (ret.) Kevin J. Sullivan, former Deputy Chief of Staff for Logistics, Installations, and Mission Support, Headquarters U.S. Air Force (AF/A4/7); Lt Gen Loren M. Reno (AF/A4/7); Lt Gen Terry L. Gabreski, Vice Commander, Air Force Materiel Command (AFMC/CV), Wright-Patterson Air Force Base, Ohio; and Michael A. Aimone (AF/A4/7), who sponsored this research and continued to support it through all phases of the project. (These and other ranks and assignments were current as of July 2009.)

On the Air Staff, we thank Maj Gen Robert H. McMahon, Director of Logistics, Office of the Deputy Chief of Staff for Logistics, Installations, and Mission Support, Headquarters U.S. Air Force (AF/A4L); Maj Gen Gary T. McCoy, formerly Director of Logistics Readiness, Office of the Deputy Chief of Staff for Logistics, Installations, and Mission Support, Headquarters U.S. Air Force (AF/A4R) and currently Commander, Air Force Global Logistics Support Center, Air Force Materiel Command (AFGLSC/CC); and Grover L. Dunn, Director of Transformation, Office of the Deputy Chief of Staff for Logistics, Installations, and Mission Support, Headquarters U.S. Air Force (AF/A4I), along with their staffs. Their comments and insights have sharpened this work and its presentation. We are grateful to our project officer, Col (S) David Koch, AF/A4LX, for his support and contributions. In addition, we would like to thank the former project officer, Col (S) Cheryl Minto, AF/A4L, for her many contributions to this effort.

At the major commands (MAJCOMs), we thank Brig Gen Kenneth D. Merchant, Director of Logistics, Headquarters Air Mobility Command (AMC/A4); and Capt Jerrymar Copeland, AMC/A4MXE, and their staffs for providing support to our analysis. At Air Force Special Operations Command (AFSOC), we would like to thank Brig Gen (S) John B. Cooper, former Director of Logistics, Headquarters AFSOC (AFSOC/A4) and Col Robert Burnett, current AFSOC/A4; Col Michael Vidal, AFSOC/A4M; Lt Col Paul Wheeless, AFSOC/A4MX; Maj Mark Ford, AFSOC/A4MO, Capt Mark Gray, AFSOC/A4MYH; and CMSgt Gerald Lautenslager, AFSOC/A4MM.

We also thank Maj Gen P. David Gillett Jr., Commander of the Oklahoma City Air Logistics Center (OC-ALC/CC). We thank Maj Gen Thomas J. Owen, former Commander of the Warner Robins Air Logistics Center (WR-ALC/CC) and then Director, Logistics and Sustainment, Headquarters Air Force Materiel Command (AFMC/A4), and Maj Gen Polly A. Peyer, current WR-ALC/CC. We also thank David Nakayama, WR-ALC Director of Staff, and Jerry Mobley, high-velocity maintenance (HVM) project lead. And we gratefully acknowledge the support of Col Richard Howard, Director of Logistics for the Air National Guard, National Guard Bureau (NGB/A4), and Col T. Glenn Davis, Director of Logistics, Headquarters Air Force Reserve Command (AFRC/A4).

We also wish to acknowledge the invaluable contributions of our RAND colleagues. Eric Peltz and David T. Orletsky provided us thorough reviews and posed thoughtful questions that substantially improved the document. Raymond A. Pyles and Brent Thomas provided us valuable insight on a number of issues.

Finally, we acknowledge the indispensable support of officer and enlisted airmen and civilians at the many bases and installations we visited. Without their support, this project could not have been done.

Abbreviations

AB	air base
AD	active duty
AF/A4/7	Deputy Chief of Staff for Logistics, Installations, and Mission Support, Headquarters U.S. Air Force
AF/A4I	Director of Transformation, Office of the Deputy Chief of Staff for Logistics, Installations, and Mission Support, Headquarters U.S. Air Force
AF/A4L	Director of Logistics, Office of the Deputy Chief of Staff for Logistics, Installations, and Mission Support, Headquarters U.S. Air Force
AF/A4R	Director of Logistics Readiness, Office of the Deputy Chief of Staff for Logistics, Installations, and Mission Support, Headquarters U.S. Air Force
AFB	Air Force base
AFGLSC/CC	Commander, Air Force Global Logistics Support Center
AFMC/CV	Vice Commander, Air Force Materiel Command

AFRC	Air Force Reserve Command
AFRC/A4	Director of Logistics, Headquarters Air Force Reserve Command
AFSOC	Air Force Special Operations Command
AFSOC/A4	Director of Logistics, Headquarters Air Force Special Operations Command
AGE	aircraft ground equipment
ALC	air logistics center
AMC/A4	Director of Logistics, Headquarters Air Mobility Command
ANG	Air National Guard
APG	Aberdeen Proving Ground
C2	command and control
CIRF	centralized intermediate repair facility
CONUS	continental United States
CRF	centralized repair facility
E&E	electrics and environment
ECM	electronic countermeasures
eLog21	Expeditionary Logistics for the 21st Century
GAC	guidance and control
HVM	high-velocity maintenance
ILM	intermediate-level maintenance
ISO	isochronal
ISR	intelligence, surveillance, and reconnaissance
JEIM	jet-engine intermediate maintenance

LCOM	Logistics Composite Model
LRU	line-replaceable unit
MAJCOM	major command
MCO	major combat operation
MDS	mission design series
NDI	nondestructive inspection
NGB/A4	Director of Logistics for the Air National Guard, National Guard Bureau
OC-ALC/CC	Commander, Oklahoma City Air Logistics Center
O-level	organizational level
OSD	Office of the Secretary of Defense
PAA	primary aircraft authorization
PAF	RAND Project AIR FORCE
PDM	programmed depot maintenance
RMF	regional maintenance facility
RNI	Repair Network Integration
SOF	special operations forces
SRU	shop-replaceable unit
TNMCS	total not-mission-capable supply
UMD	unit-manning document
UTC	unit-type code
WR-ALC/CC	Commander, Warner Robins Air Logistics Center

CHAPTER ONE

Introduction

Motivation

Planning guidance from the Office of the Secretary of Defense (OSD) directs the services to plan for high levels of engagement and deployed operations but does not specify the nature of operations, their locations, durations, or intensity. The OSD guidance depicts a world characterized by frequent global deployments. It also directs the services to develop capabilities to meet the requirements of major combat operations. At the same time, all the military services are under pressure to operate more efficiently and to meet mission responsibilities and contribute fair shares to joint taskings in support of Afghanistan and Iraq. Continuous operations have placed serious stresses on several logistics career fields, including security forces and civil engineering. Therefore, the Air Force logistics leadership is interested in finding more-efficient ways of supporting continuous aircraft deployments with fewer people. If it could be more efficient while providing the same or higher level of effectiveness—e.g., greater aircraft availability—then some of the people freed up by more-efficient support could be reprogrammed into career fields that need it most, thereby making the Air Force more expeditionary.

The Air Force logistics structure was developed primarily to support the Cold War and to meet the requirements of large-scale combat operations. This structure provided for largely self-sufficient units that carried with them significant maintenance capabilities, stocks, and other resources, on the assumption that they would be cut off from transportation for long periods.

However, the Cold War logistics support structure may not be the best one for meeting many of the demands of current and likely future requirements. The Cold War structure was tailored to support full-squadron deployments to a set of known locations and a specific operational tempo. However, more-recent engagements have called for different deployment concepts, such as those that employ only parts of squadrons and those that deploy forces to unplanned locations and for unknown durations. These partial-squadron deployments are referred to as *split operations* because they split a squadron into smaller elements to deploy. Split-squadron operations require more maintenance personnel because the squadron is operating at two locations. These additional personnel exceed authorizations, and the Air Force has decided not to fund the additional spaces. So, more-efficient ways are required to support split operations. If these more-efficient maintenance options could meet the mission requirements with fewer people than the current methods, then some of the freed-up authorizations might be provided to other career fields or applied to support split operations.

Analytic Scope

In the spring of 2007, the Deputy Chief of Staff for Logistics, Installations, and Mission Support and the Vice Commander of Air Force Materiel Command asked RAND Project AIR FORCE (PAF) to work with them and other leaders in the Air Force logistics community to refine and to develop further the vision for the Air Force logistics enterprise by designing and evaluating a set of specific logistics enterprise options that could better meet future security demands. In part, the rationale for this request was the recognition by the logistics leadership that gaps exist between the current logistics system and the vision for the logistics enterprise of the future. The Air Force has devoted significant time, effort, and money to developing transformation initiatives, largely under the Expeditionary Logistics for the 21st

Century (eLog21) program,[1] that aim at improving effectiveness and efficiency of the logistics enterprise. So, additionally, PAF was asked to carry out an independent analysis of the initiatives and ascertain whether the initiatives were appropriate for achieving the objectives of the logistics enterprise of the future.

While PAF analyses of the logistics enterprise encompass much more than maintenance processes, policies, and posture and include logistics command and control (C2), information systems and decision-support systems, inventory stockage policies, and so on, this monograph addresses only our research on developing specific enterprise options for the repair system within the logistics enterprise. The development of the repair enterprise falls under the Repair Network Integration (RNI) program within eLog21. The RNI concept calls for establishment of mission-generation units that would provide aircraft launch-and-recover and broken-component remove-and-replace maintenance capabilities. All other back-shop maintenance (that is, away from the flight line) would be performed by the logistics enterprise. Two of the recognized issues in RNI included (1) developing the criteria for allocating specific maintenance capabilities between mission-generation units and the repair enterprise and (2) evaluating the cost-effectiveness of alternative repair-network designs, from fully centralized to fully decentralized. This analysis considers ways to address these issues. More specifically, it deals with evaluating the effectiveness and efficiency of maintenance options that are capable of meeting the requirements of the new defense environment. These maintenance options address what capabilities must reside in the operating units and what can be provided from the larger logistics enterprise—independent of how flight-line maintenance is organized.

This project builds on work in the logistics area that the Air Force has sponsored over many years. Some of these projects include the following:

---

[1] eLog21 is an umbrella strategy that integrates and governs logistics transformation initiatives to ensure that the warfighter receives the right support at the right place and the right time. The eLog21 transformation campaign promotes data sharing, collaboration, and better decisionmaking across the entire Air Force supply chain. The overall goals of eLog21 are to increase equipment availability and reduce operational and support costs (USAF, undated).

- logistics concepts for rapid force deployments and employments
- regional and continental United States (CONUS) centralized intermediate repair facilities (CIRFs)
- global war-reserve materiel storage and distribution
- combat-support C2.

Our prior recommendations to use CIRFs are in the process of being implemented, but that work had limited scope. We undertook the CIRF analysis to ascertain whether centralization could provide increased maintenance efficiency (compared with traditional, decentralized structures) without reducing combat-support capability, and we focused on selected fighter engines, targeting pods, and avionics components (see McGarvey et al., 2008). That analysis showed that centralized maintenance networks outperformed decentralized maintenance networks in terms of weapon-system availability and cost in every instance except one (F-15 avionics) and that the personnel savings more than offset increased transportation costs. The effort reported in this monograph focuses much more broadly. It examines most of the maintenance processes performed at the wing level, which include sortie launch and recovery actions, removal and replacement of failed components, phase inspections, isochronal (ISO) inspections, and component repair.

Purpose

This monograph answers the following questions:

- How might the future security environment affect demands for air forces and support capabilities?
- What implications will these demands have for the strategic design of the Air Force logistics enterprise?
- What maintenance capabilities should operational units have?
- What capabilities should a global logistics network provide?
- How flexible, robust, and efficient are alternative designs?

Two points warrant mention. First, operations drive this analysis, and the first thing that needs to be examined is the operational environment that the Air Force will likely face in the future. From that environment, it is possible to derive likely demands for each aircraft mission design series (MDS) and ascertain what those demands imply for the future maintenance system.

Second, as mentioned earlier, this monograph concentrates on determining what maintenance capabilities the operational unit must have in order to meet the needs of likely future scenarios and then what capabilities can be provided from the larger logistics enterprise. We evaluate the effectiveness and efficiency of options as well as how robust they are against various kinds of risks. We also explore the possibility of additional efficiencies that might be obtained by coordinating overlapping but stovepiped processes.

Approach

Figure 1.1 shows our analytic approach. As shown in the first block, we examine the likely operating scenarios and derive maintenance workloads for each MDS that we examine. Next, as shown in the second block, we determine how that maintenance workload should be allocated among the operating units and the larger repair network supporting those units. The third step evaluates options for determining the location, the size, and the scope of centralized repair facilities (CRFs). In the fourth step, we evaluate the efficiency and effectiveness of the options. Since the repair-network design has numerous solution sets, we examine how several alternative solutions affect cost and risks, which may cause us to modify our design options. Finally, we explore the potential for additional efficiencies by synchronizing overlapping but uncoordinated processes.

Figure 1.1
Analytic Approach

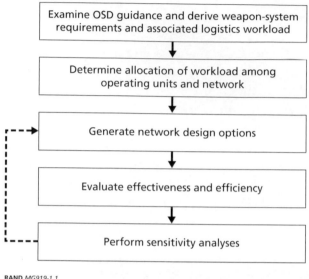

RAND *MG919-1.1*

Organization of This Monograph

The remainder of this monograph is organized in the following way. Chapter Two describes how we determined the workload and how we determined the most efficient way of allocating that workload between the operating units and the logistical network. It also explains how we measured efficiency. Chapter Three recounts the number, size, and location of the CRFs needed to absorb the maintenance workload that was allocated to them and how we determined that structure. Chapter Four explains the source of the effectiveness gains and discusses the potential additional gains that can accrue from integrating processes that currently flow in vertical stovepipes. Chapter Five presents our conclusions and outlines the future directions of the project.

CHAPTER TWO

Requirement Determination and Allocation

In this chapter, we first briefly describe the Air Force maintenance system and an example of how it operates. Those familiar with how the Air Force organizes its maintenance may wish to skip this discussion. Next, we explain how we determined the requirements for weapon systems and what those requirements imply for the logistical workload. We then show how we apportioned that workload between the operating units and the rest of the logistical network. We next describe how we determined whether the concept of shifting back-shop work to centralized facilities is more efficient than the current system and the rationale underlying that explanation.

Air Force Maintenance Practice

The Air Force generally maintains a weapon system by organizing maintenance tasks and functions into three distinct levels or echelons. In this context, *maintenance* includes the inspection, fueling, arming, and servicing of aircraft, as well as the repairing and overhauling of aircraft, its components, and associated support equipment. *On-equipment maintenance*, as the name implies, consists of maintenance work that is accomplished on the aircraft itself, while *off-equipment maintenance* refers to work accomplished on components that have been physically removed from the aircraft. The three levels of maintenance are *organizational*, *intermediate*, and *depot*.

Organizational-level (O-level) *maintenance* consists of on-equipment servicing and repair of an aircraft that is normally conducted on the flight line. An O-level repair action normally begins by identifying a failed aircraft component or line-replaceable unit (LRU), an aircraft subassembly that flight-line maintenance personnel are authorized to remove. The LRU is removed and replaced with a working spare component, and the aircraft is returned to mission-capable status.

Intermediate-level maintenance (ILM) includes both off-equipment component repair and on-equipment aircraft inspections. Off-equipment ILM consists of the repair of failed LRUs that have been removed from the aircraft, with the repair accomplished in a shop or on a test bench. Each air base establishes ILM facilities, or back shops, which are authorized to repair LRUs through the removal and replacement of failed shop-replaceable units (SRUs) or by other repair processes. The LRUs repaired through this process are then returned to the base's spare-part inventory. Each base is authorized a specific quantity of spare LRUs and SRUs to support this repair-cycle activity. Aircraft inspections, such as isochronal inspections (which are based on the elapsed time interval since the last inspection) and phase inspections (which are based on cumulative flying hours accrued on the aircraft since the last inspection) are also generally considered to be ILM activities, since these actions are typically performed in maintenance facilities dedicated to these inspections and not on the flight line.

The third level of maintenance is *depot level*. Depot-level maintenance consists of the major overhaul of aircraft through programmed depot maintenance (PDM) as well as the repair or overhaul of LRUs and SRUs. For any given aircraft or component, depot-level maintenance is usually accomplished at one central location. This location is typically an Air Force Materiel Command air logistics center (or depot), a contractor facility, or, in some cases, a Navy or Army logistics facility.

As an example of this three-level process, most air bases have a jet-engine intermediate maintenance (JEIM) facility, or engine shop. When a pilot reports an engine problem, O-level maintainers diagnose the problem. They may be able to make a minor on-equipment repair that resolves the problem. If not, they remove the engine and replace it with a serviceable spare engine. The unserviceable engine is sent to the JEIM facility, where it is inspected and disassembled. Repair is normally accomplished by removal and replacement of a major subassembly (SRU), such as a fan or compressor section. The engine is then reassembled, inspected, tested, and returned to the base's spare-engine pool. The failed SRU (in this example, the compressor) is usually returned to the depot to be overhauled or rebuilt (Lynch et al., 2007).

Determining Weapon-System Requirements and Workload

In this analysis, we derive our logistics workloads and system requirements from OSD planning scenarios. OSD instructs the services to ensure that their programs produce the capabilities to meet the demands generated by these scenarios. They call for differing MDS deployment demands in steady-state operations—e.g., notionally 10 percent of F-16 fleet, 60 percent of specialty C-130s—and they require capabilities to support major combat operations (MCOs)—that is, conflict between major combat units, typically at the national level, including from units that are already engaged in operations.

By examining the guidance in detail, we conclude that operational units, particularly in a time of limited resources, need to be as lean as possible so that they can deploy quickly to locales that may not be known in advance and so that they can be redeployed rapidly, as necessary, from one location to another. To lighten deployment burden, heavy maintenance, such as inspection and back-shop capabilities, could be provided by a centralized repair network within the logistics enterprise that can support those units with serviceable components and that can inspect and repair aircraft away from the deployed locations.

A detailed review of the guidance also shows that the steady-state burden differs for each MDS. For the fighter and bomber communities, deployments are not posited to occur as frequently as for mobility aircraft; intelligence, surveillance, and reconnaissance (ISR) assets; and special operations forces (SOF). As an illustration, fighter units may expect to deploy only 10 percent of their aircraft constantly, whereas

SOF units can anticipate deploying 60 percent of their aircraft constantly. The implication for the design of the logistics enterprise and repair network is clear: To meet steady-state requirements, not all MDS need to have the same priority. All MDS must be able to meet their MCO requirements, but, in the steady state, the Air Force has limited resources and may choose to discriminate in terms of platforms, providing those with greater demands more resources and more-robust unit-maintenance capabilities than others.

Component-repair workloads are driven directly by aircraft operating tempo—i.e., flying hours. Aircraft inspection requirements vary across weapon systems. Depending on the particular variant (known as the *block number*), F-16 aircraft require a phase inspection every 300 or 400 flying hours. For the KC-135, periodic inspections are performed when an aircraft meets the earlier of 15 months or 1,500 flying hours since its last inspection. C-130 ISO inspections take place independently of flying hours and occur 450 days after the previous inspection.[1]

Allocating Workload Between Unit and Repair Network

We next discuss how we went about determining the most-efficient allocation of workload between the operating units and the repair network that stands behind those units.

Deriving the minimum essential maintenance capability needed at operating units is not a trivial exercise. To help us come up with these unit-repair capabilities, we drew on the Logistics Composite Model (LCOM).[2] LCOM is a detailed simulation model that identifies the effect of logistics resources (primarily maintenance personnel, equipment, facilities, and spare parts) on sortie generation. We define

---

[1] We obtained these data by contacting U.S. Air Force subject-matter experts for these aircraft series.

[2] RAND developed the LCOM model 45 years ago. It has been updated frequently, and the Air Force uses it as the basis for estimating and justifying its maintenance-personnel requirements.

the minimum essential maintenance capability at the operational unit to be that necessary to perform only launch-and-recover and remove-and-replace operations. We assume that the repair network would provide the remainder of maintenance capabilities.

We use the C-130 to illustrate how we divide the maintenance capabilities between the unit and the network (see Table 2.1). We used

Table 2.1
C-130 Work Centers

Operational Unit Work Centers	CRF Work Centers
Flight-line crew chief	Aero repair, CRF
Flight-line communication and navigation	APG Inspection, CRF
Flight-line ECM	Fuels, CRF
Flight-line GAC	Metal technology, CRF
Flight-line propulsion	Structural repair, CRF
Flight-line pneudraulics	NDI, CRF
Flight-line E&E	Wheel/tire, CRF
Aero repair, unit	Communication/navigation
APG inspection, unit	ECM
Fuels, unit	GAC
Metal technology, unit	Propulsion[a]
Structural repair, unit	Pneudraulics
NDI, unit	E&E
Wheel/tire, unit	AGE
AGE	
Munitions[a]	

NOTE: Shaded cells indicate significant amount of workload in that category. APG = Aberdeen Proving Ground. ECM = electronic countermeasures. GAC = guidance and control. NDI = nondestructive inspection. E&E = electrics and environment. AGE = aircraft ground equipment.

[a] Not examined.

LCOM to compute the distribution of each work center's workload across the categories of mission generation (launch-and-recover and remove-and-replace) compared with CRF (component repair and ISO-related) workloads. The workload of many work centers falls into one category or the other. The ones that do primarily flight-line mission-generation maintenance must remain with the operational unit. The ones that do component repair and ISO-related work could be removed in their entirety and assigned to a CRF.

If component repair and ISO-related tasks are reassigned to a CRF, these work centers will be needed at the CRF location. However, because these work centers also perform mission-generation support, it would also be necessary to retain some fraction of the work-center capability at the aircraft operating location.[3] So, determining the personnel requirements at operational units and CRFs is not just a simple matter of moving all back-shop operations to a centralized facility. The shops shaded in the table must be split between the two organizations.

Evaluating Efficiency

We turn now to the evaluation of efficiency of the CRFs. We begin by explaining why we would expect to gain efficiencies with consolidation, and we follow that explanation by demonstrating the effects of consolidating C-130 facilities.

Economies of Scale

We anticipate that CRFs would be more efficient because they enjoy the advantage of labor economies of scale, which occur because larger maintenance operations can use personnel more efficiently.

Smaller, decentralized maintenance operations have relatively low personnel utilization for two reasons. One is minimum–crew size effects, which occur because the work-center task that demands the most crew to perform determines the minimum crew that can

---

[3]   Some specialty aircraft have unique work centers at the unit (e.g., armament for AC-130 gunships), and they are not presented in Table 2.1.

be assigned to the facility. This number is assigned even if most tasks performed there can be done with a smaller crew. Think of these as opening-the-door costs. A second reason is insurance effects that take into account the fact that the maintenance organization needs the capacity to accommodate random spikes in demand without too great of an adverse effect on flying operations. Conversely, centralized maintenance operations with high workload volume are able to achieve higher personnel utilization, for the same two reasons.

Consider an example drawn from our analysis of C-130 aircraft support. At a facility that performs 13 ISO inspections per year, the aero repair shop employs a minimum crew of six, which has a total direct labor utilization of 7 percent. A 65-ISO-per-year facility may also have a crew of six, but its utilization rate climbs to 34 percent. A simple linear extrapolation from 13 to 65 ISOs would suggest a crew of 30 to do the work associated with 65 ISOs (6 × [65/13] = 30), compared with the six who can actually do the work. The fact that six are assigned to the facility with many fewer ISOs is entirely attributable to minimum–crew size effects.

As the workload increases beyond 65 ISOs per year, at some point, the personnel utilization would increase to such a level that the total aero-repair personnel requirement would need to exceed the minimum crew size.

Further economy-of-scale savings can still be achieved, beyond those that result from minimum crew size. To illustrate with another C-130 example, the structural repair shop at the 13-ISO-per-year regional maintenance facility (RMF) has a simulated requirement of five maintainers; the minimum crew in this shop is three. At the 65-ISO-per-year facility, the simulated requirement is 21. The saving here ends up being four people, a much smaller number than we saw in aero repair. Because this shop exceeds its minimum crew size at both RMFs, these savings are attributable entirely to other than minimum–crew size economy-of-scale effects, such as those that result from dampening variation in demand.

These additional reductions arise because the pooling of demands dampens the effect of variations in demand, reducing the insurance premium that small facilities pay to accommodate spikes in demand.

Due to the random fluctuations associated with both the failure process and the duration of maintenance activities, personnel utilization at small maintenance operations must be kept at fairly low levels (less than 20 percent for back shops supporting F-16 squadrons of 24 primary aircraft authorization [PAA],[4] according to LCOM analysis), independent of minimum–crew size effects, to ensure that adequate capacity is available to avoid the buildup of significant queues due to spikes in demand or repair durations. As demands are pooled, back shops that are supporting ten such squadrons can meet the same level of performance (measured in terms of sortie success rate, total not-mission-capable supply [TNMCS], maintenance production rate) at a maximum personnel utilization of 45 percent as computed by LCOM.[5]

Figure 2.1 helps explain the nature of labor economies of scale. The figure plots the normalized personnel level required to perform an ISO inspection per year (the vertical axis) against the size of the CRF in terms of how many ISO inspections it can do per year (the horizontal axis). The curve and the specific points along it result from LCOM runs for our reconfigured CRF shops described in Table 2.1. The figure shows that small CRF facilities—those that conduct between 12 and 24 ISO inspections per year—require between two and three times as much manpower per ISO as large facilities that conduct 200 or more ISO inspections annually.

The Air Force has a good deal of experience in using LCOM to size manpower for facilities represented in the upper-left-hand part of this curve, but the remaining projections fall outside the range in which the Air Force has had experience with the LCOM model. Thus, it is fair to ask whether the economies of scale reflected on the right part of the curve are reasonable. As it happens, the Air Force has large facilities that perform ISO inspections. One is at Little Rock AFB. This facil-

---

[4]   PAA is the number of aircraft authorized to a unit to carry out its operational mission. The authorization forms the basis for the allocation of resources, including personnel, support equipment, and flying-hour funds.

[5]   This is driven by the desired level of performance (measured in terms of sortie success rate, TNMCS, and maintenance production rate) as computed by LCOM. For a different aircraft type and a different amount of workload (e.g., 300 C-130s), LCOM might compute a different maximum utilization achievable.

Figure 2.1
Labor Economies of Scale for Centralized Repair Facility Isochronal Inspections

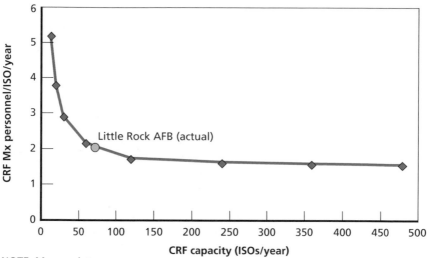

NOTE: Mx = maintenance.

RAND *MG919-2.1*

ity performs about 75 ISO inspections per year; thus, an observation based on experience shows that the economies of scale at the knee (see circle on curve in Figure 2.1) of the LCOM labor economies-of-scale curve may be reasonable expectations. Furthermore, Little Rock AFB does not use LCOM as a predictive model for determining personnel requirements. Little Rock is a training base, and its staff uses different techniques to determine the personnel required to perform ISO inspections. Thus, its personnel requirement is derived independently of LCOM but still accords with it well. The upshot is that the LCOM labor economies of scale for large CRFs are significant—reductions on the order of two or three to one—and the Air Force has some actual experience with large facilities that yield these labor economies of scale. We acknowledge that the points on the curve to the right of the Little Rock data point derive from the LCOM model and are not anchored to empirical data. However, since the Little Rock data tend to support the labor savings close to the knee of the curve, with small additional reductions beyond this point, we believe that the curve presents a reasonable estimate of the savings.

Both the LCOM data and the actual experience at Little Rock AFB show that the economies of scale occur, but they do not explain *why* they occur. The explanation appears in Figure 2.2. This chart depicts the results of LCOM runs that show that larger facilities achieve higher labor utilization of personnel than do smaller facilities.

The reasons for the increased efficiency are those outlined already—i.e., the effects of staffing crews to minimum size at small installations, the insurance premium paid by small installations, the ability of large installations to get higher utilization of personnel, and the smoothing of demands. At around 50–55 percent, rates for large facilities are relatively high. That is because constraints in LCOM limit the utilization of personnel to allow time for people to keep up with required training, ensure that tool sets are in good order, and so forth.

Having determined what minimum essential maintenance capabilities need to be associated with operational units and those that can be separated and performed at CRFs, we want to evaluate whether such

Figure 2.2
Labor Utilization Rates

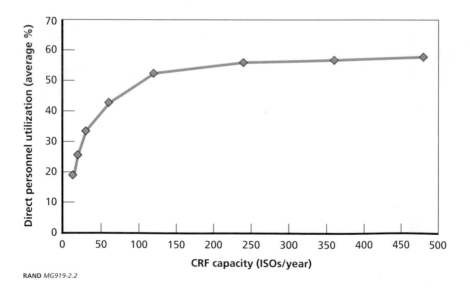

a concept is more efficient and more effective than the current system. To do this, we reconfigured the network in LCOM by establishing CRFs, and, using the C-130, we ran LCOM again to determine unit-level staffing requirements and then summed them across the fleet to obtain total system unit-level staffing requirements. We also explored a variety of CRF configurations using LCOM to determine CRF staffing requirements.

In the next two figures, we show the effects of the new personnel requirements, first at the unit level and then at the CRF. Figure 2.3 shows the results of our reconfigured unit-level LCOM runs and compares them with the current system for all active-duty (AD) and Air Force Reserve Command (AFRC) units.[6] The upper bar shows the current staffing as identified in the 22 C-130 AD/AFRC unit-manning documents (UMDs), which stipulate the personnel positions that a given unit is authorized. The left segment of the upper bar shows the staffing that is associated with maintenance-wing supervisory tasks, and the next segment shows the personnel required for munitions and JEIM. The middle portion of the bar shows current staffing associated with flight-line shops. The right segment shows the current back-shop staffing.

The lower bar shows the results of our rebalanced LCOM unit-level runs. Note that the supervision and flight-line personnel levels remain approximately the same as in the current system as do munitions and JEIM (which were not rebalanced in our analysis). The figure shows that the newly configured and rebalanced units must retain about 40 percent of the personnel spaces associated with the back shops to perform launch-and-recover and remove-and-replace maintenance. In total, the rebalanced system requires about 4,100 fewer personnel authorizations at the unit level. Note that we include both full- and part-time work in the analysis.

---

[6] Congress imposed a number of restrictions on considering consolidation involving Air National Guard (ANG) units to include certification by the Secretary of Defense that consolidation was in the national interest and that it would not harm recruiting and retention in the ANG. (See Section 324 of H. Res. 5658, 2008.) Therefore, we indicated what the savings would be consolidating only AFRC and AD units while also showing what savings would accrue if the total force were factored into centralization.

Figure 2.3
C-130 Unit-Level Active-Duty and Reserve Personnel Requirements

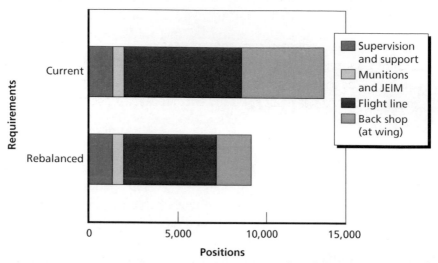

SOURCE: The data and subsequent graph were generated by RAND.
RAND *MG919-2.3*

But before it is possible to determine whether the new maintenance design is more efficient than the current system, it is necessary
to know how much manpower is required to perform ISO inspections
and component repairs at the CRFs.

Figure 2.4 shows that it takes about 1,600 CRF positions to perform ISO inspections and component repair to support all of the work
that was previously done by each wing (the rightmost segment of the
lower bar). Thus, the same work can be performed with about 2,500
fewer authorizations than the current system.

Thus far, we have determined the personnel requirements for the
new, rebalanced operational units to resource them for launch-and-
recover and remove-and-replace operations. We have also shown that
the CRFs can conduct ISO inspections and component-repair operations with much less labor than the current system because of the labor
economies of scale. In the next chapter, we turn our attention to determining how many CRFs are required, where they should be located,
and how many personnel they should have.

**Figure 2.4
C-130 Unit and Centralized Repair Facility Active-Duty and Reserve
Personnel Requirements**

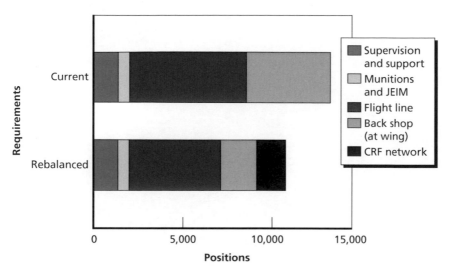

SOURCE: The data and subsequent graph were generated by RAND.
RAND MG919-2.4

CHAPTER THREE

Number, Location, and Size of Centralized Repair Facilities

Determining the numbers, sizes, and locations of CRFs is a complex mathematical problem.[1] The specific technique we employ is mixed-integer programming. This technique falls under the general heading of linear programming, and industry widely uses this approach to decide where to locate plants.

This optimization model considers trade-offs among personnel, transportation, and facility costs. The more that personnel and facilities dominate costs, economies of scale lead the algorithm and the solution technique to seek alternatives that have fewer and fewer facilities.[2] On the other hand, if transportation costs dominate, the technique will pursue solutions with more CRFs. Key parameters we use to derive the optimal solution for C-130 aircraft are as follows:[3]

---

[1] Techniques for addressing this problem were developed at RAND by George Dantzig some 50 years ago (see, e.g., Dantzig, 1956).

[2] We computed transportation and inventory requirements associated with the removal of CRF work centers from the aircraft operating locations and found them to be relatively small.

[3] We use the word *optimal* to describe the cost-minimum solution of the integer-programming model. The objective of the model is to satisfy the demand for ISO inspections in the network, given facility-, personnel-, and transit-cost parameters for a set of network locations. The Air Force may consider additional external factors in the design of the network that lead to solutions that are not optimal from the modeling standpoint but that satisfy a broader set of requirements and objectives of the Air Force (e.g., placing a CRF at an air logistics center [ALC] location). In these cases, the model can determine the cost penalty of other-than-optimal solutions allowing the decisionmaker to determine whether external factors are worth the additional cost.

- programmed flying hours and ISO intervals
- total aircraft inventory and beddown
- personnel costs ($65,000 per person-year)
- aircraft shuttle cost ($5,300 per flying hour)
- facility costs ($2 million per year per ISO dock, amortized).

With respect to CRFs, our analysis considers every potential combination, from fully decentralized solutions (that is, no CRFs and each location has the maintenance capabilities necessary to carry out maintenance tasks) to fully centralized ones. However, because of scale economies that accrue from centralization as well as the higher utilization of facilities from two- and three-shift operations at CRFs, centralized network designs are more cost-effective than noncentralized ones. For this reason, the model we use in this analysis selects centralized CRF designs.

The optimal CRF solution for the C-130 AD/AFRC repair network identified by the optimization model is shown in the leftmost bar of Figure 3.1. The optimal solution locates one CRF at Little Rock AFB to meet all AD/AFRC worldwide ISO and back-shop demands at an annual cost of $88 million. The bottom segment of the bar—the labor cost associated with the optimal solution—is the cost representation for the CRF part of the total personnel bar shown in Chapter Two—roughly 1,000 authorizations.[4] The middle portion of the bar represents the transportation costs that are incurred from shuttling aircraft from their operational location to the CRF where ISO inspections would take place and back to the operating location following the ISO. In this analysis, we charged the full price of flying those airplanes to the CRF. We assumed that they were not used for training pilots or moving cargo. This conservative assumption most likely overstates the transportation costs associated with the new CRF network because some fraction of the flights would likely be used for training and, perhaps less likely, for moving cargo. Taking existing facilities

---

[4] The CRF requires about 1,600 spaces; however, some 560 of these are allocated against work for nongrounding failures and component repair, which are independent of the network structure. Thus, we estimate cost for about 1,000 positions.

Figure 3.1
Optimized C-130 Centralized Repair Facility Solutions

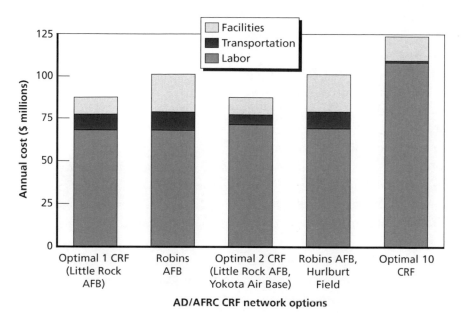

NOTE: The percentages above the least-cost solution for the other options are as follows: Robins, 115 percent; optimal 2, 100.03 percent; Robins and Hurlburt, 116 percent; optimal 10, 141 percent. Given the precision of our data, solutions within about 10 percent should be considered equal-cost options.

into account, we determined whether the selected CRF location would have to build facilities required to perform the ISO inspections. In some cases, existing facilities could accommodate the CRF. However, in other locations, facilities would have to be upgraded or constructed, and that cost is figured into the location's overall cost. Facility costs also vary across solutions because the amount of facility space required depends on the number of locations utilized (e.g., a single-CRF solution can use its space three shifts per day, while sites in a ten-CRF solution will likely only have sufficient demand to utilize their facilities one or two shifts per day).

Suppose that we wished to force the solution to be a different location, such as Robins AFB, because it might facilitate integrating stovepiped intermediate processes and depot processes.[5] The second bar in Figure 3.1 shows that the one-CRF solution is relatively insensitive to the specific location. Personnel costs remain the same. The transportation costs increase somewhat because a large number of aircraft located at Little Rock AFB would have to be moved to Robins AFB for maintenance, and there is some difference between Robins and Little Rock in terms of the transport costs per move. The facility costs for the Robins solution are larger because it is assumed that its facilities are currently fully utilized to conduct depot maintenance operations. Thus, the annual cost for the Robins solution is about $14 million more than the Little Rock one, but these costs may be more than offset if integrated processes can be achieved at Robins. We revisit this topic in Chapter Four, in which we discuss efficiencies gained from integrated processes.

Alternatively, suppose that it is deemed that one CRF for the C-130 is just too risky and the Air Force leadership wishes to have two in case something happens to one. The third bar from the left in Figure 3.1 shows the results of an analysis that forces the optimization algorithm to choose two facilities. The least-cost solution for two CRFs places them at Little Rock AFB and Yokota Air Base (AB). As shown, the personnel cost increases a little due to moving away from the maximum economies of scale, the facility costs are about the same, and transportation costs go down a little for almost an equal-cost solution.

The fourth bar in this figure shows what would happen if the Air Force chose to force an ALC into the solution—in this case, to perform all ISO inspections for the slick (i.e., unarmed) C-130s—and later pursue integrated process development. The model chooses Hurlburt Field as a second solution, and the concept would be for the Hurlburt CRF to perform ISO inspections for all SOF and weather aircraft. This solution is comparable in cost to the Robins-only solution (about $14 million) and has facility costs that are comparable for the same reasons. Thus, the Air Force would have cost latitude in where it locates the CRFs.

---

[5] Locating a CRF at an ALC site can facilitate the reengineering of those processes to include ISO inspections, phase inspections, or both.

Figure 3.1 shows that costs increase as more CRFs are added to the network. The greater the number of CRFs, the larger the personnel costs become because of diminishing labor economies of scale. The relatively small differences in costs among different numbers of CRFs and locations accord the Air Force considerable flexibility in choosing the specific CRF network. The algorithms and techniques applied here can inform Air Force CRF implementation strategies.

Comparing Results of Multiple Mission-Design Series

We have now identified the cost of alternative networks. Figure 3.2 shows the total costs for the C-130 AD/AFRC network and units. Recall from Figure 3.1 that the optimal CRF solution is in the $90 million range. We now couple that with the rebalanced unit cost and compare that with the current system cost. The left bar of Figure 3.2 shows the current system cost. The narrow band that appears on the top of this bar reflects a shuttle cost that occurs because Air Force Special Opera-

Figure 3.2
Current System Compared with C-130 Centralized Repair Facilities

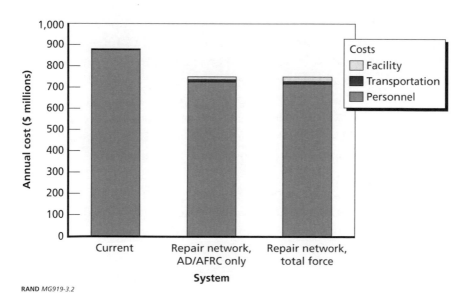

tions Command (AFSOC) has established a CRF at Hurlburt to conduct ISO inspections for most of its special-purpose C-130s. We discuss this in more depth in our discussion of high-velocity maintenance in Chapter Four. Considering the AD/AFRC solution, the middle bar, the total cost for the optimal solution is about $130 million less annually than the current one. Applying this concept to the total force network, including the ANG, the total system cost is also about $130 million less (third bar). Although there are substantial additional personnel reductions in this option, the cost savings are largely offset by increases in facility and transportation costs.

The major point that emerges from this analysis is that a rebalanced unit and CRF repair-network posture is more efficient than the current system because it can do the same amount of work with fewer authorizations, which translates into lower costs. For the AD/AFRC solution, the same work can be performed with about 2,500 fewer personnel authorizations. If this approach is applied across the total force, including the ANG, it requires about 3,200 fewer authorizations.

The relative contribution of ANG freed-up positions in the C-130 total-force CRF network is not as large as what we observed in our analyses of other aircraft (see results for the F-16 and KC-135 below). This occurs for two reasons. First, the ratio of unit-type code (UTC)[6] requirements to UMD is relatively large for ANG C-130s, which suggests that current ANG C-130 manpower lies closer to its deployment requirement, thus allowing less potential for savings. Second, additional CRF manpower necessary to support ANG C-130s is relatively large. This suggests that C-130 CRF operations afford less relative economy-of-scale personnel savings, per PAA added to the CRF network, than do KC-135 CRF operations (adding fewer PAAs causes a larger personnel addition for the C-130).

Due to the combination of these two factors, adding the ANG to the CRF network generates an additional reduction of only 636 of 3,152 personnel positions. When facility and transportation costs are included, the total cost of a C-130 total-force CRF network appears to

---

[6] A UTC is a five-character alphanumeric code that indicates a unit having common characteristics.

be essentially the same as the AD/AFRC network. However, because the ANG traditionally staffs the ISO process with one-shift operations, a large number of ANG aircraft are down for ISO-related maintenance under the current structure. As we discuss in Chapter Four, by implementing the CRF concept for the total force, the number of aircraft that would no longer be in the ISO process almost doubles in comparison to the AD/AFRC CRF network at an equal total system cost. These are large effectiveness gains, considering that the ANG aircraft represent only one-third of the C-130 fleet.

As Figure 3.3 indicates, we get similar results when we evaluate KC-135 network options. As was the case for the C-130, the optimal solution is also relatively insensitive to particular locations for CRFs, so that an ALC location can be forced into the solution with only a small cost penalty. In terms of personnel authorizations, the AD/AFRC CRF network requires about 1,100 fewer spaces to accomplish the same amount of work as the current system. The total annual cost for the AD/AFRC CRF network is about $40 million less than the current system. For the total force, the CRF network posture requires about

Figure 3.3
Current System Compared with Centralized KC-135 Repair Facilities

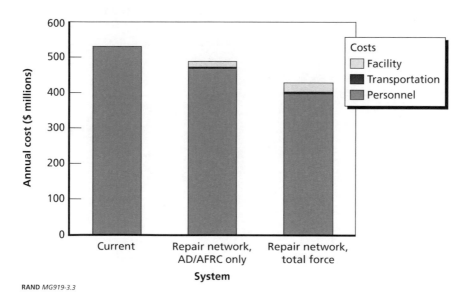

2,800 fewer authorizations and achieves an annual cost reduction of approximately $100 million when compared to the current system.

As shown in Figure 3.4, the F-16 analysis yields similar results. However, because the F-16 phase inspections take place every 300 or 400 flying hours, depending on the block designation, the network solution involves CRFs that deploy. In other words, our analysis of the steady-state security environment indicates that some deployments will be of sufficient size, duration, and intensity so that it will be necessary to set up a CRF at the deployed location. Our analyses provide assets within the fixed CRF network to support the rotation pool necessary to support such CRFs, both in the steady state and during MCOs.

The F-16 CRF supporting the AD/AFRC force can be expected to provide maintenance services with 700 fewer personnel positions for the AD/AFRC forces and about 1,800 fewer positions for the CRF to support the total force. Figure 3.4 shows that, when contrasted with the current system, the total annual cost for the AD/AFRC CRF network is about $40 million less, while the total-force network costs approximately $90 million less.

Figure 3.4
Current System Compared with F-16 Centralized Repair Facilities

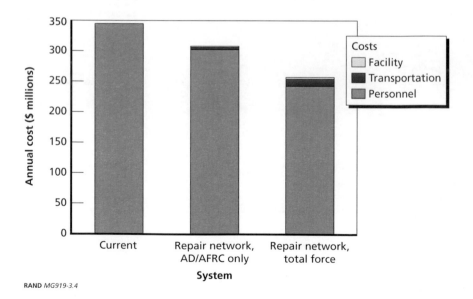

Thus far, we have shown that CRFs can perform the same workload as the current system with far fewer people. It is a point of interest that AFSOC has recognized this and has already implemented the concept. AFSOC operates some aircraft with specialized capabilities that are in high demand. However, it does not have very many of these aircraft, and they are operating at a very high rate. Additionally, AFSOC deploys these aircraft in relatively small packets—two or three at each location—which has the same effect as split operations in creating diseconomies of scale. It has implemented a centralized ISO concept, under which all of its aircraft (except those in the Pacific) return to Hurlburt Field to undergo their ISO inspections. A contractor, L-3 Communications, operates the AFSOC ISO program. AFSOC took the blue-suit personnel authorizations allocated for conducting ISO inspections and redistributed them—a rebalancing—into the operational units so that they could keep up with their operational tempo.[7]

Staffing Squadrons for Split Operations

Split operations divide a squadron's assets to support steady-state deployments while part of the squadron remains at home station. This is the way the Air Force has supported operations for more than a decade, and it is consistent with OSD guidance for the future to be prepared for continual deployments. Supporting split operations requires more supervision and results in less-efficient use of manpower due to smaller shops.

Because split-operation requirements exceed the total savings for the three weapon systems that we have examined in the AD/ARFC network, the Air Force cannot rebalance resources *and* meet all split-operation requirements. But perhaps the Air Force does not need to resource all MDS for split operations. It might decide to distribute the savings differentially by MDS. OSD programming guidance indicates that some forces—SOF, ISR, and mobility—might be more heavily taxed in steady-state operations than others are. Therefore, the Air

---

[7] *Blue suit* refers to uniformed Air Force.

Force might consider taking some of the manpower freed up by centralizing back-shop operations and provide more unit-level manpower for these forces and not provide additional resources to meet split-operation requirements for fighters and bombers. Thus, the Air Force could resource various MDS and units on a differential basis because their steady-state deployment may not be equal. Alternatively, the Air Force might apply some of these savings freed up from maintenance to pay larger Air Force bills, including addressing personnel shortages in other areas, such as combat support, force protection, civil engineers, or those associated with numbered Air Force staffs.

We used LCOM to identify the additional manpower requirement associated with staffing for split operations. Table 3.1 shows a total annual requirement for split operations of $250 million and a savings from the three aircraft series we analyzed of only $207 million.

Table 3.1
Requirements for AD/AFRC Split Operations

Aircraft Type	Potential CRF Network Savings		Split-Operation Annual Requirement	
	($ millions)	Spaces	($ millions)	Spaces
F-16	37	720	53	844
KC-135	43	1,092	69	1,213
C-130	127	2,516	128	2,395
Total	207	4,328	250	4,452

CHAPTER FOUR

Effectiveness Analysis

To this point, the discussion has dealt with the efficiencies generated by CRFs. We turn now to the effectiveness advantages of the CRF network solutions. Here we show that consolidation not only yields efficiency gains but also enables more-effective utilization of aircraft.

Source of Effectivness

Consolidation of workload through centralization can make more aircraft available because it can reduce the time that it takes aircraft to move through the inspection process, resulting in more available aircraft. Figure 4.1 shows the expected flow time, on the vertical axis in days, as a function of the size of the centralized facility in terms of its capacity for ISO inspections.

The points on this curve were generated from LCOM. The figure shows that the flow time (i.e., the time it takes to conduct an inspection) is longer for small facilities than it is for large ones. The number of people and the number of shifts employed in the operation affect the time it takes to inspect an aircraft. Smaller facilities typically operate with one or two shifts per day, usually one with significant staffing and one with light staffing. Moving to the right on the curve, the larger facilities employ two- and three-shift operations. Thus, the larger labor pool at larger facilities enables more-effective operations that decrease flow times.

Since most of the Air Force's experience in managing ISO operations confines itself largely to the left portion of this curve, should we

Figure 4.1
Effect of Facility Size on Inspection Time

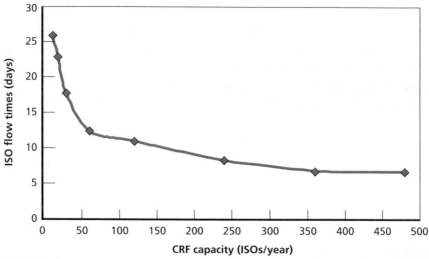

have confidence that the reduction in flow days, associated with large facilities in the right region of the curve, is feasible? Again, the Air Force does have some experience with facilities operating at the knee of this curve. Figure 4.2 shows that AFSOC, which does about 75 ISO inspections per year, has ISO flow in the 12-day range. This facility runs three shifts per day, seven days per week. Little Rock also does about 75 ISO inspections per year, but it operates two full shifts per day with a skeleton crew on the third shift. So again, the Air Force's experience with large consolidated facilities shows that the LCOM model runs produce credible results in terms of what may be expected in ISO flow times, at least up to 75 ISO inspections.

Figure 4.3 shows how the reduction in flow days associated with centralized ISO inspections can significantly reduce the number of aircraft tied up in ISO and refurbishment processes and make more aircraft available for operational use. As shown in this figure, the current system has about 53 airplanes occupied in that process. Moving to a C-130 CRF network that supports the AD/AFRC fleet, the number of aircraft in the ISO process would drop to about 34, making almost 20 additional aircraft available to the operational

Figure 4.2
AFSOC and Little Rock AFB ISO-Inspection Flow Times

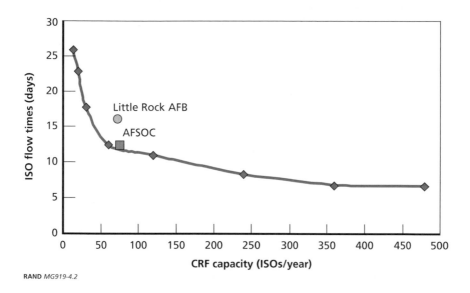

RAND *MG919-4.2*

Figure 4.3
Number of C-130 Aircraft in ISO Inspection, as a Function of Network Type

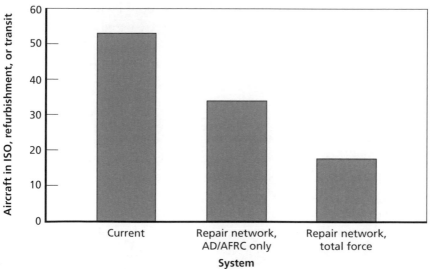

RAND *MG919-4.3*

units. If the network were to support the total force, another 17 aircraft would become available for operational use.

Integrating Processes

A final point to consider is that consolidation, done wisely, can be a stepping stone to integrating stovepiped maintenance processes that exist across the enterprise. For such aircraft as the C-130, the unit-level ISO process and the depot-level PDM process are often not coordinated. Several elements of the ISO inspection, including removing panels and inspecting specific areas of the aircraft, are identical to tasks performed during the PDM process at an ALC. However, the ISO and PDM processes are often not aligned due to the calendar inspection intervals or because units do not want to incur the additional costs if the inspections are performed at the ALC. There are several examples of aircraft that return to a unit from a PDM and immediately undergo the ISO process. From the viewpoint of the enterprise, these redundant tasks waste maintenance manpower and increase aircraft downtime. By integrating maintenance processes, such as ISOs and PDMs, even greater efficiencies and effectiveness benefits could be realized.

All of our analyses in this monograph were based on current processes, as defined in the LCOM models. We showed that there was not much cost penalty if an ALC were forced into the set of chosen CRF solutions compared with the optimal solution set determined by the optimization model. Moreover, if an ALC were a CRF site, where current PDM or modification operations occur, the site could facilitate the reengineering of the processes to include synchronizing ISOs or phase inspections with a PDM or modification program. Engineers can study those operations and work to improve the system continuously and provide greater efficiency and effectiveness.

An example of an integrated maintenance concept is being developed for the C-130 fleet at Warner-Robins ALC under the name of high-velocity maintenance (HVM). HVM divides the PDM process into four stages, each performed at the depot. The existing calendar for the ISO inspection process will be synchronized with HVM, meaning that each time an aircraft visits the depot for a portion of a PDM, an ISO is performed. HVM will also leverage two- and three-shift operations to speed up the flow time through the process and greatly decrease the average number of aircraft in the maintenance process compared to the sum total of the PDM and ISO processes. We note that the workload of the HMV process does not necessarily need to be consolidated only at the ALCs. In fact, we have shown that there are ranges of options within the vicinity of the optimal solutions and locations that can be selected for a variety of reasons.

CHAPTER FIVE

Conclusions

Our major overarching conclusion is that consolidated wing-level scheduled inspections and component back-shop maintenance capabilities would be more effective and efficient than the current system, in which every wing has significant maintenance capabilities to accomplish these activities. Consolidation yields efficiencies because it requires fewer people. It is more effective because consolidation can speed the flow of aircraft through inspections, which means that fewer aircraft are tied up in maintenance processes at any given time and, thus, more aircraft are available to the operational community. Consolidating scheduled inspection and back-shop operations not only would provide immediate benefits but also could provide a good basis for integrating stovepiped intermediate- and depot-level processes, thereby creating the opportunity for even greater efficiencies and effectiveness.

This analysis also shows that the Air Force has considerable flexibility in locating CRFs with respect to achieving projected cost savings. Selecting an ALC as a CRF site would facilitate the reengineering of current PDM or modification processes to include ISO and phase inspections without necessarily incurring a cost penalty. Furthermore, locating a CRF at a depot can offer some other benefits. Engineers can study those operations and have a laboratory to use in transforming those processes to provide greater efficiency and effectiveness, as illustrated by the HVM example at Warner-Robins ALC. This does not mean that all workload needs to be consolidated at only the ALCs. In fact, we have shown that there are ranges of options within the optimal solutions that can be selected to minimize risk or for other reasons.

Bibliography

Clay, Shenita L., *KC-135 Logistics Composite Model (LCOM) Final Report, Peacetime and Wartime, Peacetime Update*, Scott AFB, Ill.: Headquarters Air Mobility Command XPMMS, May 1, 1999.

Dantzig, George Bernard, *Thoughts of Linear Programming and Automation*, Santa Monica, Calif.: RAND Corporation, P-824, 1956.

Directorate of Technical Support, Air Force Civil Engineer Support Agency, *Historical Air Force Construction Cost Handbook*, Tyndall AFB, Fla., February 2004. As of August 21, 2009:
http://www.afcesa.af.mil/shared/media/document/AFD-070613-072.pdf

H. Res. 5658—*see* U.S. House of Representatives (2008).

Hoehn, Andrew R., Adam Grissom, David A. Ochmanek, David A. Shlapak, and Alan J. Vick, *A New Division of Labor: Meeting America's Security Challenges Beyond Iraq*, Santa Monica, Calif.: RAND Corporation, MG-499-AF, 2007. As of August 21, 2009:
http://www.rand.org/pubs/monographs/MG499/

International Air Transport Associates, "Jet Fuel Price Monitor," undated Web page. As of July 30, 2008:
http://www.iata.org/whatwedo/economics/fuel_monitor/

Lynch, Kristin F., John G. Drew, Sally Sleeper, William A. Williams, James M. Masters, Louis Luangkesorn, Robert S. Tripp, Dahlia S. Lichter, and Charles Robert Roll Jr., *Supporting the Future Total Force: A Methodology for Evaluating Potential Air National Guard Mission Assignments*, Santa Monica, Calif.: RAND Corporation, MG-539-AF, 2007. As of August 21, 2009:
http://www.rand.org/pubs/monographs/MG539/

McGarvey, Ronald G., Manuel Carrillo, Douglas C. Cato Jr., John G. Drew, Thomas Lang, Kristin F. Lynch, Amy L. Maletic, Hugh G. Massey, James M. Masters, Raymond A. Pyles, Ricardo Sanchez, Jerry M. Sollinger, Brent Thomas, Robert S. Tripp, and Ben D. Van Roo, *Analysis of the Air Force Logistics Enterprise: Evaluation of Global Repair Network Options for Supporting the F-16 and KC-135*, Santa Monica, Calif.: RAND Corporation, MG-872-AF, 2009. As of January 11, 2010:
http://www.rand.org/pubs/monographs/MG872/

McGarvey, Ronald G., James M. Masters, Louis Luangkesorn, Stephen Sheehy, John G. Drew, Robert Kerchner, Ben D. Van Roo, and Charles Robert Roll Jr., *Supporting Air and Space Expeditionary Forces: Analysis of CONUS Centralized Intermediate Repair Facilities*, Santa Monica, Calif.: RAND Corporation, MG-418-AF, 2008. As of August 21, 2009:
http://www.rand.org/pubs/monographs/MG418/

Public Law 108-325, Ronald W. Reagan National Defense Authorization Act for Fiscal Year 2005, October 28, 2004. As of August 21, 2009:
http://www.dod.mil/dodgc/olc/docs/PL108-375.pdf

U.S. Air Force, "Expeditionary Logistics for the 21st Century," undated Web page. As of August 21, 2009:
http://www.af.mil/information/elog21/

―――, *KC-135 Logistics Composite Model (LCOM) Final Report: Peacetime and Wartime*, Scott AFB, Ill.: Headquarters Air Mobility Command XPMMS, May 1, 1995.

―――, *Statement of F-16 Block 40 Aircraft Maintenance and Munitions Manpower*, Langley AFB, Va.: Headquarters Air Combat Command XPM, August 2003.

―――, "Expeditionary Logistics for the 21st Century," Washington, D.C., modified March 25, 2005. As of August 27, 2009:
https://acc.dau.mil/CommunityBrowser.aspx?id=22406

―――, *US Air Force Cost and Planning Factors, BY 2008 Logistics Cost Factors*, Washington, D.C.: Air Force Cost Analysis Agency, Air Force Instruction 65-503, Annex 4-1, February 22, 2006.

―――, *US Air Force Cost and Planning Factors, FY2008 Standard Composite Rates by Grade*, Washington, D.C.: Secretary of the Air Force for Financial Matters, Air Force Instruction 65-503, Annex 19-2, April 2007a.

―――, Deputy Chief of Staff for Installations and Logistics Directorate of Transformation, *Logistics Enterprise Architecture (LogEa) Concept Of Operations*, May 24, 2007b.

―――, Table of Distribution and Allowance Personnel Detail metadata, November 2007c.

U.S. Department of the Air Force, Future Concepts and Transformation Division, Deputy Chief of Staff for Plans and Programs, *Air Force Transformation Flight Plan*, Washington, D.C., PBD 720, December 2005.

U.S. Department of Defense, *Quadrennial Defense Review Report*, Washington, D.C., September 30, 2001. As of August 21, 2009:
http://purl.access.gpo.gov/GPO/LPS18834

————, *Base Realignment and Closure Commission Action Brief*, September 1, 2005.

U.S. Department of Defense and Joint Staff, *Mobility Capabilities Study*, Washington, D.C., September 19, 2005. Not available to the general public.

U.S. House of Representatives, House Resolution 5658, 110th Congress, Duncan Hunter National Defense Authorization Act for Fiscal Year 2009, introduced March 31, 2008. As of August 21, 2009:
http://thomas.loc.gov/cgi-bin/bdquery/z?d110:h.r.05658:

U.S. House of Representatives Committee on Armed Services Panel on Roles and Missions, *Initial Perspectives*, Washington, D.C., January 2008. As of August 21, 2009:
http://purl.access.gpo.gov/GPO/LPS95952

USAF—*see* U.S. Air Force.

Van Roo, Ben D., Manuel Carrillo, John G. Drew, Thomas Lang, Amy L. Maletic, Hugh G. Massey, James M. Masters, Ronald G. McGarvey, Jerry M. Sollinger, Brent Thomas, and Robert S. Tripp, *Analysis of the Air Force Logistics Enterprise: Evaluation of Global Repair Network Options for Supporting the C-130*, Santa Monica, Calif.: RAND Corporation, forthcoming.